T0012207

THE LITTLE BOOK OF

THE GRAND CANYON

Published in 2023 by OH!
An Imprint of Welbeck Non-Fiction Limited,
part of Welbeck Publishing Group.
Offices in: London – 20 Mortimer Street, London W1T 3JW
and Sydney – 205 Commonwealth Street, Surry Hills 2010
www.welbeckpublishing.com

ISBN 978-1-80069-390-6

Compiled and written by: Malcolm Croft
Editorial: Victoria Denne
Project manager: Russell Porter
Design: Tony Seddon
Production: Jess Brisley

A CIP catalogue record for this book is available from the British Library

Printed in China

10 9 8 7 6 5 4 3 2 1

THE LITTLE BOOK OF

THE GRAND CANYON

A BREATH-TAKING EXPERIENCE

CONTENTS

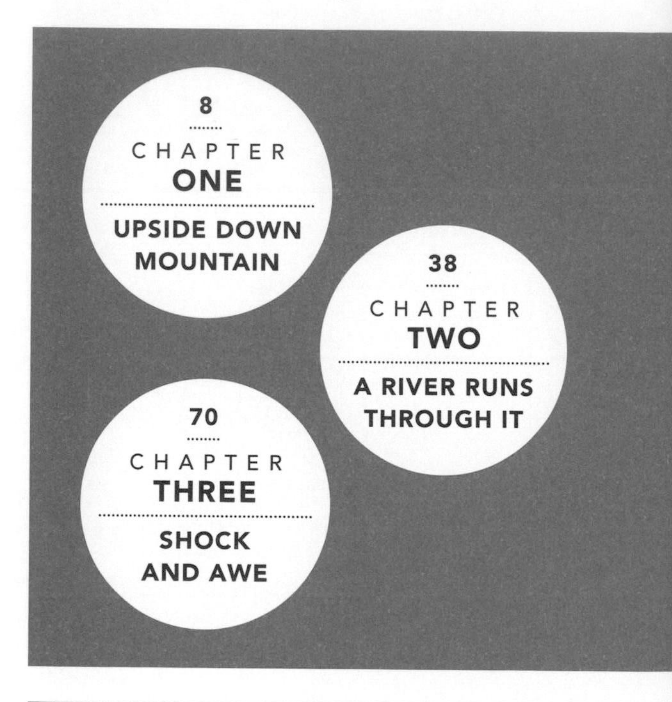

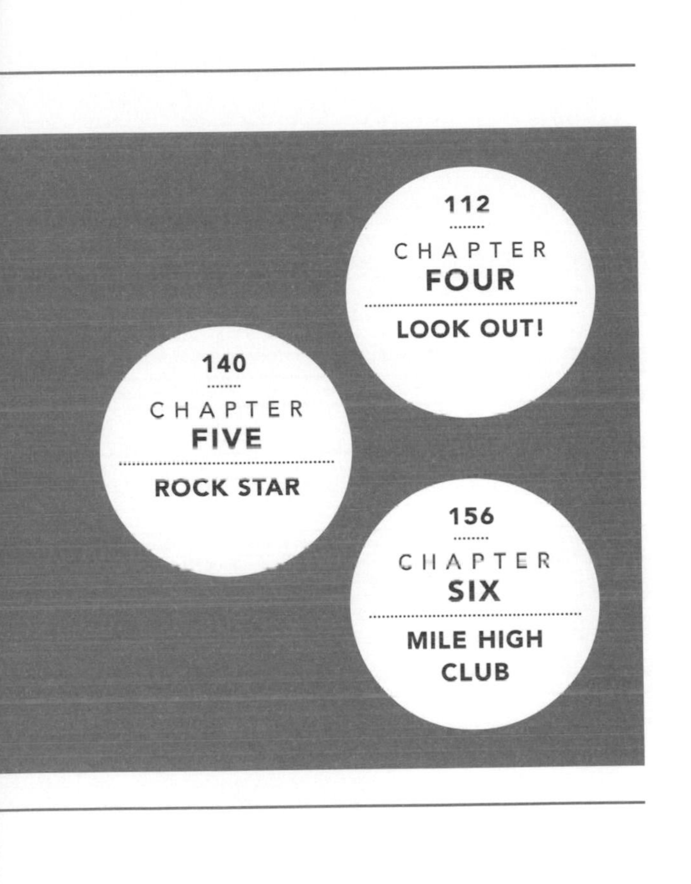

INTRODUCTION

Welcome to our little celebration of the very large, and very lovely, Grand Canyon! As its name suggests, the Grand Canyon is very grand indeed. Though why it isn't called the Jumbo Canyon or the Colossal Canyon, or the Humongous Canyon, is beyond me. That makes it sound much grander. Anyway, this mega-massive monument to Mother Nature's might is simply jaw-dropping, so get ready to bend down a lot to pick it up.

The enduring appeal of this natural masterpiece, and what defines its legendary status beyond its impressive size, is its never-ending beauty, of course. I like to think of it as the place where Mother Nature comes to boast her inner beauty.

From the stacked pancake rock strata that shines a bright spotlight on 1.8 billion years of geological history to its iconic waterway – and founding father – the Colorado River, with its world-famous white-water rapids, to its host, the Grand Canyon National Park, and its sacred Native American ancestry. There is so much more to the Canyon than its looks. It's deep, man, like well spiritual too. It is a place before, and beyond, time.

As is so often observed, the Grand Canyon is an icon of America, but in many ways, it is America. The perfect representation of all that is loved about the United States. It's big, it's bold, it's beautiful.

Of course, you no doubt already know how big and how beauty-filled this craggy crack is. Any postcard (or selfie) from the edge can tell you that. What you don't know about the Grand Canyon, however, well, may well fill the Grand Canyon, to borrow that dog-eared phrase. And that's where this tiny tome comes in…

While the Grand Canyon may be visible from space, this compendium of all things canyon-esque is so compact it'll fit snugly in your glovebox or backpack, so you can take it with you as you blaze your next canyon trail. Whatever this mini pick-'n'-mix of trivia may lack in size, it makes up for in heavy-hitting stats, facts, quotes and quips – and companionship too, I like to think – filled to the brim with all the awesome bitesize brain snacks you'll need to keep your quest to the Grand Canyon as jam-packed with action and fact-ion as possible. If you leave home without it, you'll get lost. It's happened before.

But enough rambling…next stop…Arizona!

CHAPTER
ONE

Upside Down Mountain

At the Grand Canyon, the view up is as epic as the view down. In fact, everywhere you look, this heaving bosom of Mother Nature's beautiful majesty is a sight for awe-struck senses unfolding as far as the eyes can see. So, let's see it all. Grab your oars, and binoculars, because it's time to take your mind, body and soul on an unforgettable journey 1.8 billion years (give or take) in the making…

The uppermost level of Canyon rock is known as "Kaibab". This creamy, white limestone is relatively young, just 230 million years old, when compared to the granite rock at the bottom of the Canyon, which is more than 1.8 billion years old. The name "Kaibab" originates from the Paiute Native American tribe. It means "mountain turned upside down", which is probably the best way to describe the Grand Canyon.

“ Crying. Acceptable at funerals and the Grand Canyon. ”

Ron Swanson, *Parks, and Recreation*, ‘Rainy Day’

5,562 trillion

The weight of the
Grand Canyon in tons.

According to the annual National Park Service report, more than 4.5 million people visited Grand Canyon National Park in 2021, a total of more than $710 million for local tourism.

This spending supported 9,300 jobs within the Grand Canyon National Park.

Rock of Ages: Thermochronology

A super-sophisticated rock-aging method called thermochronology uses heat to tell the timelessness, and age, of the Grand Canyon. Rocks, when stacked on top of each other under immense pressure, create a lot of heat. The rocks begin to "toast". This toasting leaves behind a chemical footprint that reveals clues as to the age of the rocks.

“

I believe in science and evolution. I’ve been to the Grand Canyon.

”

Bill Walton

“Ooh, it’s like Disneyland for thin people.”

Marge Simpson, *The Simpsons*, ‘Fland Canyon’

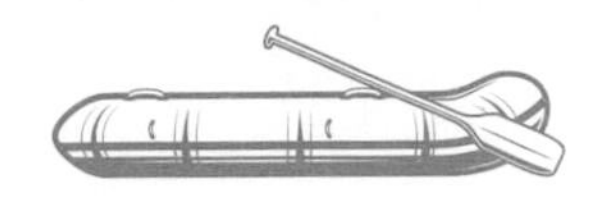

The first great (white) icon of the Canyon is John Wesley Powell. It was he who first mapped the Colorado River in 1891 on a wooden boat.

Powell was also the very first person to use the name “Grand Canyon”, changing it from “Big Canyon’ in 1871.

Genius!

1 in 400,000

Odds of dying at the Grand Canyon.

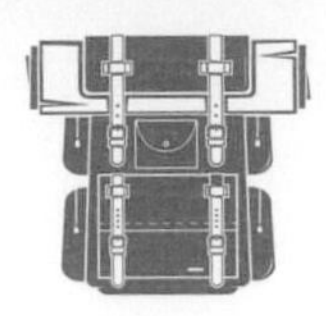

“Jesus, it’s only the biggest damn hole in the world.”

Clark Griswold, *National Lampoon’s Vacation*

On average, about 12 people die per year in the Grand Canyon.

The leading cause of death is airplane and helicopter crashes, falling, and environmental causes, such as heat stroke, drowning and dehydration.

Out of that 12, two to three deaths per year are from falls over the rim, most commonly from posing for a photo, or selfie.

According to *Conde Nast Traveller*, 57 per cent of Americans don't know what state the Grand Canyon resides in.

15 per cent

Percentage of Americans alive yet to visit the Grand Canyon.

The Grand Canyon is home to five separate ecosystems, so pack accordingly:

1. Boreal Forest
2. Ponderosa Pine Forest
3. Pinyon-Juniper Woodland
4. Desert Scrub
5. Desert

1,000 calories an hour

A Grand Canyon hiker can burn as many calories per hour as a marathon runner. Don't forget to pack your mixed nuts.

Astronauts flying high in outer space on the International Space Station are some of the lucky few who get to view all 277 miles of the Grand Canyon all at once. It remains one of the few landmarks visible from space, along with these locations:

1. The Great Pyramids of Giza, Egypt
2. The Himalayas
3. Great Barrier Reef, Australia
4. Amazon River
5. Palm Island, Dubai
6. Ganges River Delta
7. Greenhouses of Almería, Spain
8. River Thames, UK
9. Kennecott Copper Mine, USA

To save you time before the sun sets on your Grand Canyon trip, here's the not-so-secret seven hotspots at the South Rim to see the sun set.

1. Mather Point
2. Hopi Point
3. Mohave and Pima Point
4. Yavapai Point
5. Lipan Point
6. Navajo View
7. Desert View

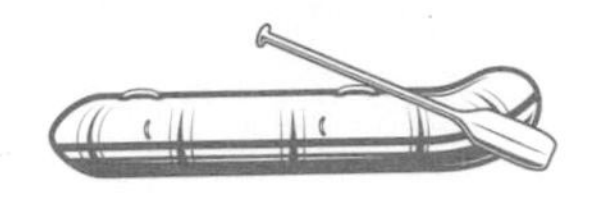

The Grand Canyon isn't the only colossal hole in the Arizona Desert.

About 110 miles away from the Grand Canyon lies Meteor Crater, the best-preserved meteorite crater on Earth. Formed 50,000 years ago when a meteor struck the earth, leaving a 3,900 ft (1,200 m) diameter hole 560 ft (170 m) deep. The impact released more energy than 20 million tons of dynamite.

When in Rome

Ten incredible places to stop en route – or on the way back – from the Grand Canyon.

1. Lake Mead
2. Meteor Crater
3. Petrified Forest National Park
4. Hoover Dam
5. Antelope Canyon
6. Route 66
7. Tombstone town
8. Horseshoe Bend
9. London Bridge
10. Apache Death Cave

The Great Unconformity

Down by the riverside, the Grand Canyon poses its most peculiar puzzle: a geological phenomenon that has baffled geologists for decades.

Known as the Great Unconformity, it is where 250-million-year-old layers of rock lie back-to-back against rocks that are 1.8 billion years old. A near-billion years' worth of rocks appear to be missing in the Canyon sides without a satisfying answer.

Can you solve the riddle?

66.2 trillion

The amount of people required to completely fill the Grand Canyon is roughly 9,000 times the amount of our current global population of 7.8 billion.

Arizona is the only state to contain all four main North American deserts – the Sonoran, the Mohave, Chihuahuan, and the Great Basin.

In 2010, McDonald's Japan released a "Grand Canyon Burger" to celebrate its Big America campaign, alongside a Las Vegas Burger, a Broadway Burger, and a Beverly Hills Burger.

The Grand Canyon Burger comprised two beef patties with a "steak filling", a cooked egg, cheddar and mozzarella cheese, crispy onions, steak sauce with soy and liquid smoke, and an extra middle bun for good measure.

Most white-water-rapid rafters are aware of the international 1–6 river classification scale for white-water rivers. However, due to the sheer power of the Colorado River's rapids within its stretch of the Grand Canyon, a scale from 1 to 10 is used instead – one of only a handful of places around the world where this is the case.

“The Grand Canyon is our cathedral in the desert, and the word ‘our’ is key because although the Canyon belongs to the entire world, we, as Americans, belong particularly to it.”

Kevin Fedarko

“Across the Colorado River from the Needles, the dark and jagged ramparts of Arizona stood up against the sky, and behind them, the huge tilted plain rising toward the backbone of the continent again.”

John Steinbeck

Fred Flintstone: "Boy, the Grand Canyon. That's one of nature's wonders. Let's take a look."

Wilma Flintstone: "It doesn't look like much to me."

Fred Flintstone: "Not now. But they expect it to be a big thing someday."

***The Flintstones*, 'Droop Along Flintstone', 1961**

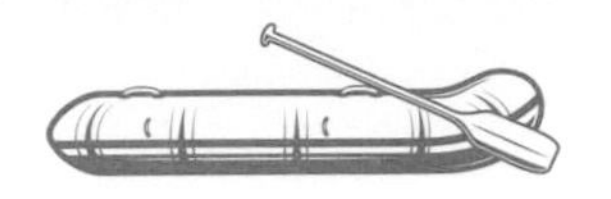

30 to 40 per cent

Annual percentage of the five million or so visitors to the Grand Canyon who are foreign tourists.

CHAPTER
TWO

A River Runs Through It

The Canyon would be nothing but a lot – *a lot* – of flat dull rock without the founding forces of the mighty 1,500-mile Colorado River, a life-giving freshwater that has spent its entire existence sculpting and snaking the Canyon into what we know and love today. All hail the king of the Canyon – the Colorado!

“

Eventually, all things merge into one, and a river runs through it. The river was cut by the world’s great flood and runs over rocks from the basement of time. On some of the rocks are timeless raindrops. Under the rocks are the words, and some of the words are theirs. I am haunted by waters.

”

Norman Maclean, *A River Runs Through It*

Top Ten Things Tourists Say When They See the Grand Canyon for the First Time

1. "Wow!"
2. "It's bigger than I expected!"
3. "I'm speechless."
4. "It's…just…so… BIG!"
5. "Well worth all the fuss."
6. "We can finally tick it off the Bucket List!"
7. "Take a photo of me, will you."
8. "I can fit it all in the photo!"
9. "This'll make John at home well jealous."
10. "I don't want to go home."

How many bananas would it take to fill the Grand Canyon?

One *really* big one.*

**We've done the maths – 26 quadrillion bananas would fill the Grand Canyon!*

“There will never be a photograph of the Grand Canyon that can adequately describe its depth, breadth, and true beauty.”

Stefanie Payne

94 per cent

The percentage of the Grand Canyon National Park considered wilderness.

“You can’t say you’re going to jump the Grand Canyon and then jump some other canyon.”

Evel Knievel

At its lowest point, the Grand Canyon is a mile (5,280 ft) deep.

That's twice as tall as the world's longest man-made structure, the Burj Khalifa skyscraper, Dubai, which tops out at 2,717 feet, (about half a mile).

Just like any unforgettable tale, the Grand Canyon has a beginning, middle and end.

It begins at Lees Ferry, in northern Arizona, 50 miles south of the Arizona/Utah border, and ends at Grand Wash Cliffs, east of Las Vegas, Nevada. Here, Lake Mead and the Hoover Dam start a whole new story of modern American history.

The Grand Canyon's one mile (5,280 ft) depth could fit 31 Nelson's Column (UK), 13 Great Pyramids of Giza (Egypt), five Eiffel Towers (France), three One World Trade Centres (USA), and two Burj Khalifas (Dubai) inside of it.

(Not all at once, obviously.)

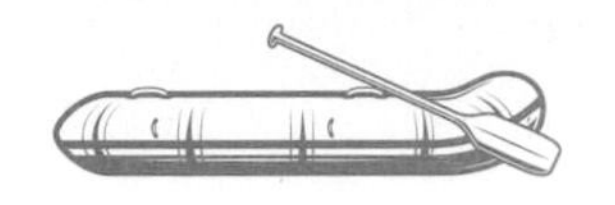

The Colorado River is the largest river in America's Southwest. This famous waterway kickstarts its 20,000-cubic feet-per-second flow in Colorado state's Rocky Mountains and drains the mountain ranges of four states (Utah, California, Nevada and Colorado) from their snow/rainwater. The river comes to an epic end after 1,450 miles (2,333 kilometres) in the Gulf of California, where frankly it deserves a rest.

“

The Grand Canyon is living evidence of the power of water over a period of time. The power may not manifest immediately. Water can be very powerful, like a tidal wave.

”

Frederick Lenz

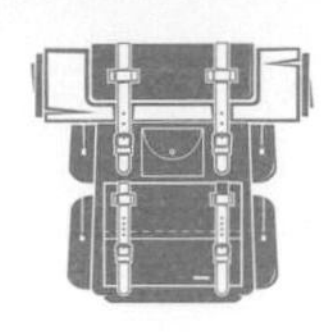

Geologists have recently confirmed that it's likely that dinosaurs, which became extinct 65 million years ago, never had the pleasure of roaming the Grand Canyon. While there is much debate about how old the Grand Canyon is, most estimates suggest it is between 7 and 17 million years, and up to 70 million years, but, to date, no dinosaur fossils have ever been found.

Welcome to Supai, population: 208

Supai Village is the smallest, most remote and isolated community in the continental United States. It is hidden inside the Grand Canyon, located eight miles from the nearest road. The only way to reach the village is by helicopter or a rather extreme eight-mile hike with only a train of mules and sheer cliff drops to keep you company. Supai is the last official place in the U.S. where post is delivered by mules.

Havasu Falls

One of the highlights of Canyon hiking is Havasu Falls, five stunning waterfalls flowing over exposed 1.8-billion-year-old rocks. Located at the foot of Hualapai Hilltop, Havasu Falls is the ancient home of the Havasupai Native American tribe, who have lived here for more than a millennia. Havasupai means "people of the blue-green waters". And quite rightly so!

The Grand Canyon's pink rattlesnake is found nowhere else in the world. It's a "super sneaky" species, blending superbly into the surrounding red-hued rocks, and are rarely seen. But, be warned, they are supremely venomous! If you see one – run.

To date, no one has died due to a snakebite in the Canyon, though several visitors have been bitten.

Joseph Christmas Ives was one of the earliest American soldiers and explorers to lead an expedition to the Grand Canyon. In 1861, Ives reported to Congress about the monetary value of the landscape and got it completely wrong:

"The region is, of course, altogether valueless. It can be approached only from the south, and after entering it there is nothing to do but leave. Ours has been the first, and will doubtless be the last, party of whites to visit this profitless locality. It seems intended by nature that the Colorado River, along the greater portion of its lonely and majestic way, shall be forever unvisited and undisturbed."

86023

The Grand Canyon has a postcode. Or ZIP* code if you prefer. Send all your fan mail here:

20 South Entrance Road, Grand Canyon, AZ, 86023.

Did you know that ZIP stands for Zonal Improvement Plan?

In 1857, Edward Fitzgerald Beale – one of the Western Frontier's first famed explorers, led an expedition to survey the Colorado River. He viewed the Canyon's charm thusly: "A wonderful canyon four-thousand feet deep.

Everyone in the party admitted that we have never before saw anything to match or equal this astonishing natural curiosity."

According to the creation stories of the Hualapai, a Native American tribe, the Grand Canyon was created after a great flood occurred in the region.

To drain the land, a giant mythical Hualapai hero used a large knife and club to cut and beat the Earth so all the flood water would flow back to the Pacific Ocean.

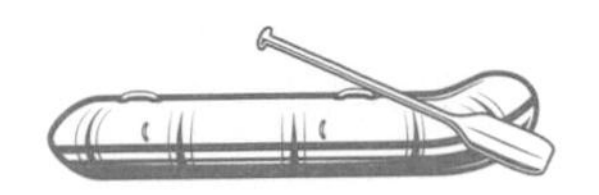

“The Grand Canyon is a land of song. Mountains of music swell in the rivers, hills of music billow in the creeks, and meadows of music murmur in the rills that ripple over the rocks. Altogether it is a symphony of multitudinous melodies.”

John Wesley Powell

On 29 November 1976, at 14:20 local time, an unrestrained German Shepherd aboard a Grand Canyon Air Piper 32-300 interfered with the cockpit controls, causing the plane to plummet and crash. Everyone on board died.

It is the first, and only, time a dog has bought down an airplane over U.S. territory.

300

The amount of Search and Rescue operations performed by Park Rangers at the Grand Canyon in 2021.

Each rescue costs the park about £500,000 dollars a year.

The National Park Service does not charge for rescues if they use their own equipment.

50

The amount of Grand Canyon National Park Rangers.

Park Rangers live and work in three main visitor areas: North Rim, South Rim, and Desert View. Outside of that, you're on your own.

“The bigger the better; in everything.”

Freddie Mercury

Ooh Aah Point

One of the Canyon's most fabulously named viewpoints, one that offers unobstructed views of the Canyon as far as the eye can see, is Ooh Aah Point.

This scenic spot, located 600 feet below the South Rim, is part of the Canyon's most popular and accessible trails – the South Kaibab – a 6.3-mile hike that descends 4,800 feet to the rapid-flowing Colorado River.

486 quadrillion pounds

That's approximately 6,817.4 quadrillion slices of bacon – just enough to fill the Grand Canyon. Though it would take a billion years to fill it, based on current pork production.

Grand Canyon Mule

To honour the hard-working mules of the Grand Canyon, we've remixed the Moscow Mule cocktail to make the perfect antidote to a day-long hike down the Canyon. Unlike your hike down, this cocktail won't touch the sides.

Serving Suggestion

50ml vodka
Crushed ice
150ml ginger beer
A dash or two of ginger bitters
Sprig of mint
Wedge of lime

Make It Right

Mix the vodka with the ginger beer and the crushed ice, then garnish with mint leaves and lime. For that extra Canyon feel, play the "Canyon Playlist" from page 148.

In 2021, after the Covid-19 pandemic, more white-water rafters paddled down the Colorado River than ever before – 619,968 commercial trips, to be precise, an increase of more than 189,793 rafters from 2019.

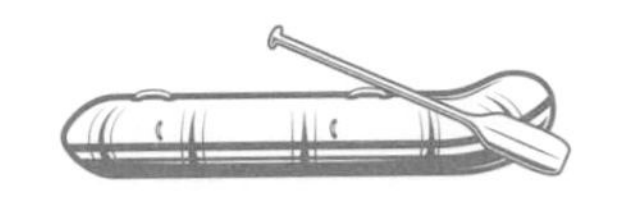

Arizona is the only state in the U.S. (besides Hawaii, obviously) that does not apply daylight saving time in the summer. The state follows Mountain Standard Time all the time that's one hour ahead of Pacific Standard Time (PST).

CHAPTER
THREE

Shock and Awe

There's more to the Grand Canyon than just the Grand Canyon. From tourism treks to wilderness trails, mules to rules, and park rangers to park dangers. Let's explore all that Mother Nature's greatest work of art has to offer in all its various shapes and sizes, one bitesize snack at a time…

“

Once you’ve been in the Canyon and once you’ve sort of fallen in love with it, it never ends. In fact, I’ve often said that if I ever had a mistress, it would be the Grand Canyon.

”

Barry Goldwater

Want to see the grandest canyon in our Solar System? Head to Valles Marineris – the largest known canyon on Mars. It spans about 20 per cent of the entire distance around Mars and is about 2,000 miles (3,000 kilometres) long, 370 miles (600 kilometres) wide and 5 miles (8 kilometres) deep.

On Earth, it would stretch from New York to California!

“The wonders of the Grand Canyon cannot be adequately represented in symbols of speech, nor by speech itself.”

John Wesley Powell

Located in the northern Arizona desert – and taking up a 1,904-square-mile chunk of it – the Grand Canyon is measured by the path of the Colorado River along the bottom of the Canyon. Today, that measures at 277 miles (446 kilometres) long, 18 miles (29 kilometres) wide and 1 mile (1.6 kilometres) deep. For your money, you get 4.17 trillion *cubic metres* of layered red rock, caves, scrub and, of course, the Colorado River.

The Grand Canyon was carved by the Colorado River over a period of more than 6 million years. The river was able to carve a crack through the Arizona desert due to the Colorado Plateau – the earth's crust – becoming raised 7,000 feet (2,133 metres) above sea level during a period of plate tectonic shift that occurred around 80 million years ago.

“

I believe in evolution. But I also believe, when I hike the Grand Canyon and see it at sunset, that the hand of God is there also.

”

John McCain

The Grand Canyon is home to many different species: more than 1,500 plants, 355 birds, 89 mammals, 47 reptiles, 9 amphibians, and 17 fish.

The Grand Canyon is the name of the largest shopping mall in northern Israel, located in the city of Haifa. The name of the mall is a play on words – “mall” in Hebrew is pronounced “canyon”. It has more than 220 stores, just FYI.

There are 11 Native American tribes that call the Grand Canyon National Park home. Know their name…

1. Havasupai Tribe
2. Hopi Tribe
3. Hualapai Tribe
4. Kaibab Band of Paiute Indians
5. Las Vegas Band of Paiute Indians
6. Moapa Band of Paiute Indians
7. Navajo Nation
8. Paiute Indian Tribe of Utah
9. San Juan Southern Paiute Tribe
10. The Pueblo of Zuni
11. Yavapai Apache Nation

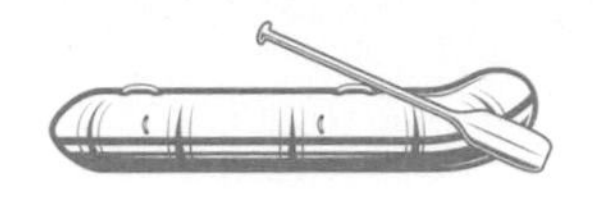

“

In the Grand Canyon, Arizona has a natural wonder which is in kind absolutely unparalleled throughout the rest of the world.

”

Theodore Roosevelt

900,000,000,000,000

(That's 900 trillion.)

That's how many footballs would fill the Grand Canyon, top to bottom.

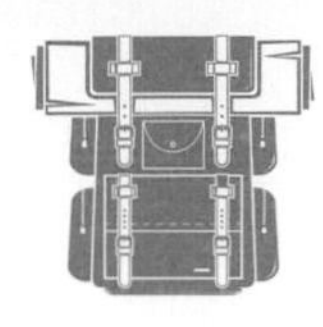

The name Colorado is borrowed from the Spanish language, meaning “reddish colour”.

The Colorado River took this name due to its colour, caused by its heavy volume of silt (sand and clay).

In 2020, at the height of the Covid-19 pandemic, trail runner Lars Arneson broke a world record when he double-crossed the Grand Canyon on the 40-mile "Rim to Rim to Rim" route in a personal best of 6 hours, 38 minutes, 42 seconds (beating the 2011 record by 15 minutes).

Arneson's 20-mile route from the South Rim to the North Rim includes a 5,000-foot descent into the Canyon, then running 7 miles across the Canyon floor and concluding with a 6,000-foot climb up the other side of the Canyon. Then back again.

Fancy it?

“

The Grand Canyon fills me with awe. It is beyond comparison – beyond description; absolutely unparalleled throughout the wide world... Let this great wonder of nature remain as it now is. Do nothing to mar its grandeur, sublimity and loveliness.

”

Theodore Roosevelt

Rock, Paper, Squirrel

The Grand Canyon is home to mountain lions, black bears, condors, charging elk, gila monsters and Bighorn Sheep. However, the most dangerous predator of them all is the… rock squirrel.

During busy holiday weekends, more than 30 visitors, on average, require medical assistance after receiving rock squirrel bites. "Enjoy squirrels from a safe distance," park officials say. "Their sharp teeth crack nuts – and cut fingers."

In 1933, a 43-year-old prospector named Cochrane, from California, who reportedly had an intense phobia of snakes, was hiking down (rather aptly) Snake Gulch (near the North Rim) when a pink rattlesnake rattled and coiled up – preparing to strike. It didn't; but Cochrane leaped backward in fear and died immediately of a heart attack, confirmed later by a doctor.

36.0544° N, 112.1401° W.

Longitude and latitude coordinates for the Grand Canyon.

In case someone asks.

Between the
2,209 miles from
Grand Cayman, in the
Caribbean, and the
Grand Canyon, Arizona,
you could squeeze in
eight Grand Canyons.

“The world is big, and I want to have a good look at it before it gets dark.”

John Muir, Father of America's National Parks

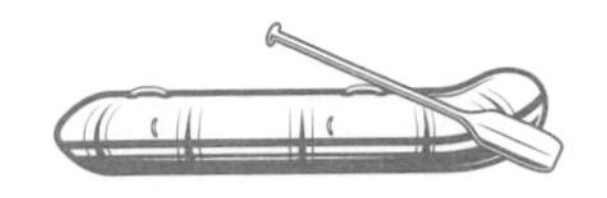

$16.78 per hour.

The average salary for a Grand Canyon Park Ranger.

Perks of the job: the view – priceless.

The Colorado River flows at an average speed of 4 miles per hour, a speed considered to be moderately fast-flowing.

In 2022, the Grand Canyon was ranked America's most Instagrammed National Park, claiming the highest amount of Instagram hashtags – 4,067,301. The other most Instagram-ideal National Parks are:

1. Acadia, Maine

2. Zion, Utah

3. Arches, Utah

4. Hot Springs, Arkansas

5. Bryce Canyon, Utah

6. Yosemite, California

7. Pinnacles, California

8. Cuyahoga Valley, Ohio

9. Rocky Mountain, Colorado

2,403,294

The amount of football pitches you could fit inside the Grand Canyon's 1,218,375 acres.

If you're an adrenalin junkie fresh from a white-knuckle white-water ride of the rapids below, and looking for your next fix, why not try the "Grand Canyon Swing Ride" to get your heart pumping again.

The swing dangles its rider 1,300 feet above the Colorado River, before propelling them 112 degrees above the horizon – at 50mph!

Must be seen to be believed.

"

To watch the morning light, illuminate the shadowed crevices of the Canyon, as the sky turns a truly astonishing blue, is to witness one of the most miraculous wonders of the world. I took my time. I took it in. And I'll keep it with me forever.

"

Oprah Winfrey

“

If there is a point to being in the Canyon, it is not to rush but to linger, suspended in a blue-and-amber haze of in-between-ness, for as long as one possibly can. To float, to drift, savouring the pulse of the river on its odyssey through the Canyon, and above all, to postpone the unwelcome and distinctly unpleasant moment when one is forced to re-emerge and re-enter the world beyond the rim – that is the paramount goal.

”

Kevin Fedarko

In 1983, Kenton Grua, Rudi Petschek and Steve Reynolds set a record for speed rowing down the entire 277-mile length of the Colorado River through the Grand Canyon, in a time of 36 hours, 38 minutes and 29 seconds.

"You're just blown away that this river could cut this canyon, that could just go on and on and on," Grua said. "On the water, the Canyon is different every day, different every minute and every second…just this unfathomable amount of water is flowing by over these rapids and flats, and the walls just keep getting higher, and it just keeps getting better and better and better and better every day, and it just seems like it's going to go on forever...and then 'Boom' it's over."

In 1977, American Kenton Grua became the first person in recorded history to hike the length of the Grand Canyon in a single trip.

The 600-mile journey took him 36 days. He had to abandon his first attempt earlier in the year after stepping on a cactus.

82.77 per cent

The UK is almost the same size as the state of Arizona. Bizarrely, the UK has almost precisely 10 times the population – 7 million in Arizona, 70 million in the UK.

*At its widest, the UK is also 300 miles wide – just 23 miles longer than the Grand Canyon.

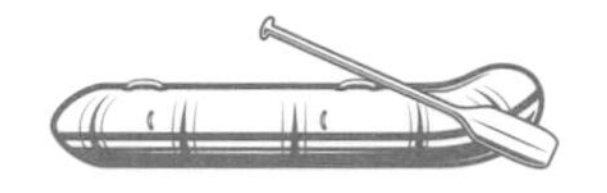

“

Are you tired of the same old Grand Canyon?

”

Tom Hanks, *The Simpsons Movie*, 2007

"

The park rangers, and probably the Secret Service behind me, got quite nervous. Sort of like my mom did the first time I went to the Grand Canyon when I was 11. I was getting a little too close to the edge.

"

Barack Obama

According to most estimates, the Grand Canyon is home to at least 80 big water rapids.

"

Tonight is very lovely,
The river is swelling by,
And high above the canyon walls
A cloud flecked, sunset sky.

"

Clyde Eddy*, 1927.

*Clyde Eddy first saw the Colorado River in 1919 and vowed to someday travel its entire length. Eight years later, he did. Forty-two days and 800 miles later, he became the first to successfully navigate the river during its annual high-water period.

With a drop of 37 feet, a water flow of 18,000 cubic feet per second, and a rapid that has the power of 75,000 horses (more than double the power output of a Triton nuclear submarine), Lava Falls Rapids is the Colorado River's mightiest Class 10 white-water rapid. It is also one of the most powerful, and hardest, sections of white water anywhere in North America. It's a beast!

The good news? The rapid is short – lasting about 20 seconds from the top to the bottom.

32 per cent

The percentage of Earth's geologic history found within the Grand Canyon.

The last ice age occurred around 11,000 to 80,000 years ago. Large mammals such as the woolly mammoth and giant sloth – which then could grow to 12 feet tall! – roamed around the Grand Canyon, according to fossils of both species discovered within the National Park.

Loch Ness, Scotland, is as wide as the Grand Canyon is deep.

However, Grand Canyon's length is 10 times that of Loch Ness.

The Hopi tribe believe that the sipapuni, or "place of emergence", is a gateway, or door, to the afterlife.

The sipapuni is a dome of mineral deposits built up from a nearby hot spring on the banks of the Little Colorado River.

The Hopi tribe believe when the spirit of the recently deceased passes over the dome it transcends the physical and enters the spiritual afterlife.

Theodore Roosevelt, the 26th president of the United States, died in January 1919, a month before his beloved Grand Canyon and National Parks became joined at the hip forever in February 1919.

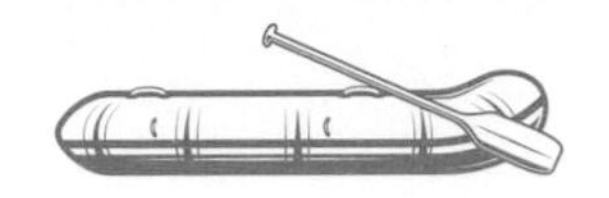

2 quadrillion gallons

Covering 1,218,375 acres (1,904 square miles), the Grand Canyon can hold a lot of water.

If you poured all the river water on Earth into the Grand Canyon, it would only ever get about half full.

Try it if you don't believe us.

CHAPTER
FOUR

Look Out!

For more than a century since the foundation of the National Park, explorers and adventurers from all over the world have come to the Grand Canyon to drink in its handsome brilliance. But, for those who come and look closer, they'll find a whole lot more than what they came looking for. It's time to mine for buried treasures…

“

This has got to be the most beautiful thing we’ve ever stole from the Indians.

”

Homer Simpson, *The Simpsons*, ‘Fland Canyon’

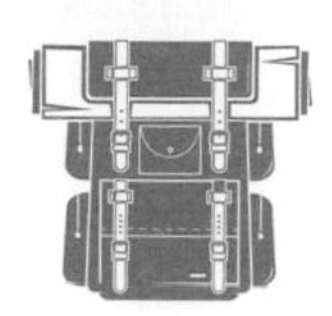

“

I can still remember my first experience of standing at the edge of the Grand Canyon and looking into it. It was so awesome, it took a fair amount of restraint to prevent me from jumping into it, because I was certain I could fly.

”

Mark Goulston

“

The glories and the beauties of form, colour, and sound unite in the Grand Canyon.

”

John Wesley Powell

The Grand Canyon National Park, as it has been known since 1919, has been continuously inhabited by several Native Americans tribes, including the Hopi, the Hualapai, the Havasupai, and the Navajo.

However, it was the ancient and deeply spiritual Anasazi people who were the original inhabitants of the Canyon and its thousands of caves, calling it "Ongtupqa", in Hopi language, which translates as "Salt Canyon".

Only 12 people have ever successfully completed a continuous 800-mile hike through the entire length of the Grand Canyon. That's the same amount of people who have walked on the moon.

Why? For 90 per cent of the Canyon there is no trail.

On average, most hikers will take five hours to reach the Canyon floor, no matter which trail they choose. However, on average, it takes most hikers another eight hours to get back up.

Pack wisely.

The Colorado River is the 47th longest river in the world. Prone to constant flooding, the Hoover Dam, built between 1931 and 1936, was constructed to control the flooding of the Colorado River, as well as provide hydroelectric power for the millions of people venturing west towards California. The water collected by the dam is called Lake Mead.

More than
40 million people
rely on the
Colorado River as
a water source.

Famously, the Grand Canyon is one of the seven wonders of the natural world.

What are the other six?

The Northern Lights, Paricutin, Mount Everest, Harbour of Rio de Janeiro, Victoria Falls and the Great Barrier Reef.

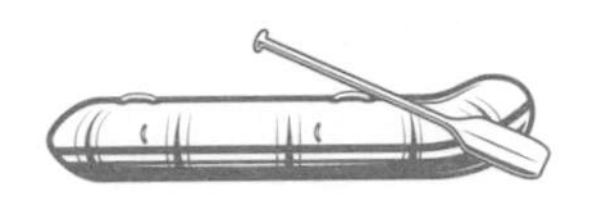

American president from 1901–1909, Theodore Roosevelt declared the Grand Canyon a national monument in 1906 through the National Monuments Act, a law that was passed to preserve the heritage of the nation.

“Keep it for your children and your children’s children, and for all who come after you, as one of the great sights which every American . . . must see,” he wrote at the time. “Leave it as it is. You cannot improve on it. The ages have been at work on it, and man can only mar it.”

Technically, the Grand Canyon is not even the grandest canyon in America. Kings Canyon, California, is the deepest canyon in the USA, at more than 8,200 feet (2,500 metres).

Oh, and there are more than 70 grand-ish canyons across North America.

The exposed granite rocks discovered at the lowest level of the Canyon have been dated at 1.8 billion years old – approximately half of the age of Planet Earth*. In fact, the rocks at Grand Canyon are some of the oldest rock types found anywhere else around the globe, including marine and volcanic.

* The oldest ever bedrock discovered can be seen in the Hudson Bay in Quebec, Canada. It dates back 4.6 billion years – as old as Earth's origins.

Grand Canyon was – finally – designated a National Park by U.S. Congress in 1919.

To date, only 13 of the 63 U.S. National Parks are UNESCO World Heritage sites:

1. Carlsbad Caverns National Park, New Mexico

2. Olympic National Park, Washington State

3. Yellowstone National Park, Wyoming

4. Mammoth Cave National Park, Kentucky

5. Glacier National Park, Montana

6. Redwood National Park, California

7. Mesa Verde National Park, Colorado

8. Hawaii Volcanoes National Park, Hawaii

9. Great Smoky Mountains National Park, Tennessee

10. Glacier Bay National Park, Alaska

11. Grand Canyon National Park, Colorado

12. Yosemite National Park, California

13. Everglades National Park, Florida

25,000

The annual amount of lightning strikes recorded with in the Grand Canyon.

From March 2012, the Grand Canyon National Park officially ceased the sale of water in plastic bottles. This was decided after a report showed that 20 per cent of the Canyon's waste came from disposable water bottles.

Approximately 53 people fell to their deaths from the Canyon rims from 1925 to 2005, with another 48 deaths inside the Canyon. On rare occasions, people have even driven their vehicles straight into the Canyon off the South Rim, Thelma and Louise style.

500,000 visitors

Due to its more remote location, open only in summer months, the North Rim receives just 10 per cent of the 5 million who visit the South Rim.

The world-famous daredevil Evel Knievel never got the chance to jump the Grand Canyon on his motorcycle. But his son Robbie did.

On May 20, 1999, Robbie jumped 228 feet (69 meters) across the Canyon, at its narrowest point. Fireworks erupted as he soared! If he had failed, Robbie would have plummeted 2,500 feet (762 meters) to his death – in front of a live television audience of tens of millions.

On March 23, 2006, Bob Burnquist skateboarded off the edge of the Grand Canyon, following an awesome 40-foot (12-metre) ramp jump!

Burnquist soared 1,500 feet (457 meters) above the Canyon floor, before opening his parachute.

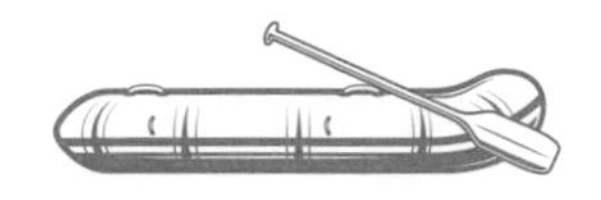

Rock Songs: Songs About Rock

1. 'Rock with You' – Michael Jackson
2. 'Rock You Like a Hurricane' – Scorpions
3. 'I Wanna Rock' – Twisted Sister
4. 'Rock the Casbah' – The Clash
5. 'Rock Around the Clock' – Bill Haley & His Comets
6. 'Loves Me Like a Rock' – Paul Simon
7. 'We Will Rock You' – Queen
8. 'Rock N Me' – Steve Miller Band
9. 'Cherub Rock' – Smashing Pumpkins
10. 'Let There Be Rock' – AC/DC

Mules have carried more than 1 million visitors in and out of the Canyon since the late 1800s. They've yet to complain.

In 1903, President Theodore Roosevelt became one of the first visitors to ride a mule down the Canyon.

“

I can’t believe I have to say this, but we can’t let Donald Trump open up the Grand Canyon for uranium mining.

”

Joe Biden

Home to the highest bat diversity in the United States, the Grand Canyon provides a habitat for 22 species.

The Little Brown Bat (*Myotis lucifugus*) can eat 1,200 insects an hour. But best to still bring repellent.

Mountain lions (*Puma concolor*) – or pumas, cougars and panthers, as they are also known – are the largest predators found in the Grand Canyon. But don't worry, they're not interested in humans.

Yet.

But they may be in the future… if food sources continue their downward trend.

Due to dramatic changes in elevation, the Grand Canyon contains five of the seven life zones that exist on earth – Lower Sonoran, Upper Sonoran, Transition, Canadian, and Hudsonian.

A life zone is a region defined by its plant and animal species; a concept developed in 1889 to classify similar regions on Earth together.

When Albert Einstein visited the Grand Canyon in 1931, members of the Hopi tribe invited the world-famous physicist to wear a traditional feather headdress. They also gave him a Hopi Peace Pipe, as a recognition of his renowned pacifism.

They also affectionately bestowed him with his own tribal name, "The Great Relative", after his gravity-defying unravelling of the universe, *The General Theory of Relativity*, published in 1915.

CHAPTER
FIVE

Rock Star

The Grand Canyon's infamous "Great Unconformity" has baffled rock stars (geologists) for decades, becoming one of the known unknowns of Earth's 4.5-billion-year history. However, there are many other strange mysteries under our feet here too. From Bigfoot to handprint pictographs, curses and conspiracies, the Canyon is bursting with secrets waiting for us to snaffle…

“

Baseball, it is said, is only a game. True. And the Grand Canyon is only a hole in Arizona. Not all holes, or games, are created equal.

”

George Will

“I wish they’d build a ski jump at the Grand Canyon; it’d be fantastic.”

Eddie the Eagle

At 277 miles (446 km), the Grand Canyon is one-fifth as long as the mighty river that runs through it – that's long enough to fit 4,875 football fields!

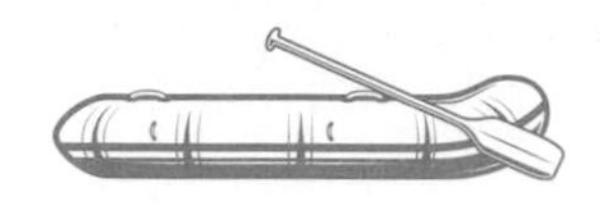

In 1919, the Grand Canyon became America's 14th National Park. Today, it is the second most visited park in the U.S. behind the Rocky Mountains.

In 2021, 300 million people (roughly the entire population the U.S.) visited at least one of the 63 National Parks across America, generating $42 billion for the U.S. economy.

12,000 years

Archaeological evidence suggests that the Anasazi tribe were the first to roam the Canyon, 12,000 years ago.

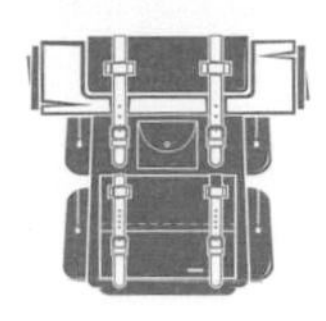

At 1.2 million acres (1,902 square miles), Grand Canyon is the 11th largest national park in the United States.

Here are the Top Ten:

1. Wrangell-St. Elias, Alaska (33,682.60 km²)

2. Gates of the Arctic, Alaska (30,448.10 km²)

3. Denali, Alaska (19,185.80 km²)

4. Katmai, Alaska (14,870.30 km²)

5. Death Valley, California (13,793.30 km²)

6. Glacier Bay, Alaska (13,044.60 km²)

7. Lake Clark, Alaska (10,602.00 km²)

8. Yellowstone, Wyoming/Montana/Idaho (8,983.20 km²)

9. Kobuk, Alaska (7,084.90 km²)

10. Everglades, Florida (6,106.50 km²)

Rock Songs: Classic Canyon Tracks

1. 'Ladies of the Canyon' – Joni Mitchell

2. 'Echoes Down the Canyon' – Cal Jones

3. 'Devil's Canyon' – Molly Hatchet

4. 'Girl from the Canyon' – Johnny Cash

5. 'Twelve-Thirty (Young Girls Are Coming to the Canyon)' – Mamas and Papas

6. 'Grand Canyon Song' – Steve Goodman

7. 'Canyon Moon' – Harry Styles

8. 'The Canyon' – Claire Heywood

9. 'Torrey Canyon' – Serge Gainsbourg

10. 'The Call of the Canyon' – Frank Sinatra

Today, the majority of the 5 million or so people who visit the Canyon drive there by car.

The first automobile to drive to the Canyon in 1902 was a Toledo Eight Horse steam car, built by the American Bicycle Co. Driven by Oliver Lippincott, the 60-mile journey from Los Angeles was expected to take three hours – it took two days. He didn't pack any food or water – only smoking tobacco and matches – and hallucinated from dehydration. When he finally reached the Canyon, Lippincott realized the trip was worth it. "I stood there upon the rim and forgot who I was and why I came there for. Before me lay the sublimest panorama in the world."

The most famous movie scene featuring the Grand Canyon is, of course, *Thelma & Louise.*

At the dramatic climax of the film, Thelma and Louise sit in their 1966 Thunderbird – surrounded by police, following their exploits, which we won't spoil here – and refuse to surrender. Instead, they vow to "keep on going" and drive straight over the Grand Canyon to their death.

“The Arizona desert takes hold of a man’s mind and shakes it.”

David W. Toll

“There are in the world valleys which are larger and a few which are deeper. There are valleys flanked by summits loftier than the palisades of the Kaibab. Still the Grand Canyon is the sublimest thing on earth. It is so not alone by virtue of its magnitude, but by virtue of the whole – its ensemble.”

Clarence Dutton

The oldest fossils discovered with in Grand Canyon are 1.2 billion years old. They are the remains of microorganisms called Stromatolites, formed by the photosynthesizing of bacteria called cyanobacteria.

They are the oldest living lifeform – and first evidence of life on Planet Earth.

Want to taste the Grand Canyon? Now you can. And, thankfully, it tastes like beer.

The Grand Canyon Brewery, established in 2007, was created to pay homage to the breath-taking landscape of the Canyon and the adventurous lifestyle of living in Williams, Arizona.

This craft beer specialist is located at 301 North 7th St, Williams, AZ 86046, and currently sells 14 types of craft ales, pilsners and IPAs named after clean Canyon living.

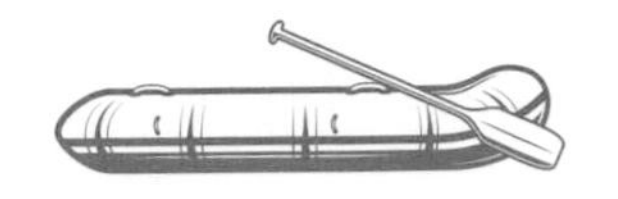

The two major cities closest to the Canyon are Phoenix and Las Vegas.

CHAPTER SIX

Mile High Club

We've reached the end of our quest to see as much of the Grand Canyon as possible and buy the t-shirt too. But wait! There's still a lot more to explore on the way home as we work out just how on Earth we get out of this mile-deep chasm in the Arizona desert – the hottest in all the U.S. Thankfully, the road home is just as fun as the road here…

"In fact, just about all the major natural attractions you find in the West – the Grand Canyon, the Badlands, the Goodlands, the Mediocrelands, the Rocky Mountains and Robert Redford – were caused by erosion."

Dave Barry

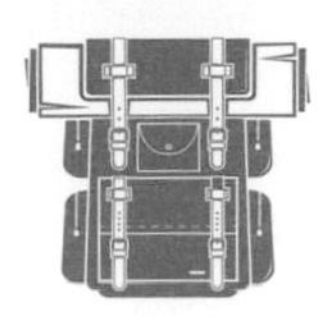

"

Nothing prepares you for the Grand Canyon. No matter how many times you read about it or see it pictured, it still takes your breath away. Your mind, unable to deal with anything on this scale, just shuts down and for many long moments you are a human vacuum, without speech or breath, but just a deep, inexpressible awe that anything on this earth could be so vast, so beautiful, so silent.

"

Bill Bryson

"When I did the video for 'Holding Out for A Hero', we filmed that on top of the Grand Canyon, and that was quite frightening. I was close to the edge, and there was a helicopter hovering about, creating a lot of wind. I was nervous I was going to fall off."

Bonnie Tyler

For about 6 million years, the Colorado River has carved a steep valley through the rock. As a result, the river has exposed 1.8 billion years of Earth's geological history, highlighting the four main geological eras of Earth: Precambrian and Proterozoic (granite), Palaeozoic (limestone) and Mesozoic (sandstone) and Cenozoic – the current geological era, representing the last 66 million years.

“

It’s like trying to describe what you feel when you’re standing on the rim of the Grand Canyon or remembering your first love or the birth of your child. You have to be there to really know what it’s like.

”

Jack Schmitt

“

The sun rises over the
Grand Canyon, igniting rocks
that have been there for two
billion years before we were
born and will likely remain two
billion years after we’re gone.
My heart aches with the cruel
and unimaginable beauty of it.
We are nothing.

”

Sarah Ockler

4.5 million

The number of visitors the Grand Canyon welcomed in 2021*. Before the pandemic, visitor numbers hit record highs of 6.3 million in 2019.

** In 1919, when the park was created, only 44,173 canyoners thought it was worth a visit.*

1,904 square miles (4,931 km²)

This is the size of the Grand Canyon. That's bigger than the entire state of Rhode Island (1,212 square miles), twice the size of Tokyo and three times the size of London.

Only 335 of the 1,000 naturally formed caves tucked away in the Canyon have been recorded. And only a minority of them have been studied. There is a wealth of Native American culture and geological history still be to be found in the Canyon's cave system.

Today, only one cave is open to the public – the Cave of the Domes on Horseshoe Mesa at the end of the Grandview Trail.

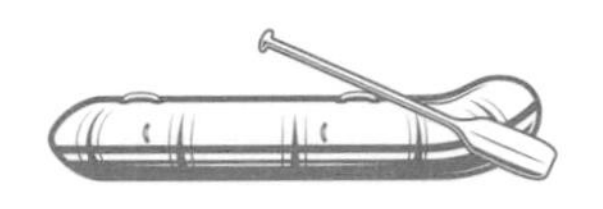

During the 2021 season, park rangers needed to hire a dozen hunters to shoot an overpopulation of bison within the park. The request received 45,000 applicants! Grand Canyon officials ask volunteers "to be skilled and serious about the operation… and every volunteer is required to pass a firearms safety course and a marksmanship proficiency test."

Volunteers were required to haul bison carcasses, which can weigh more than 2,000 pounds, by hand and foot, out of the park, without the aid of a vehicle.

The Canyon's two most scenic hotspots – the North Rim and South Rim – are only 10 miles apart, as the crow flies. However, visitors wishing to see both rims on the same day will be required to undertake a 215-mile journey around the entire Canyon top – a trip that takes more than five hours by car.

There are actually a few bridges across the Grand Canyon at its least wide points: the Navajo Bridge (277 metres), the Black Bridge (130 metres) and the Silver Bridge (150 metres).

The Silver Bridge connects the South Rim of the Grand Canyon to Phantom Ranch and the North Rim.

More than 800,000 people fly over the Grand Canyon in helicopter and airplane tours every year.

In 2007, the Grand Canyon Skywalk was unveiled to the world.

This horseshoe-shaped steel walkway, with its acrophobia-inducing glass-bottomed floor, extends 25 metres (70 feet) out from the Canyon rim. As terrifying as that sounds, incredibly, the Skywalk's five 2.5-inch-thick layers of glass can support the weight of 71 fully loaded 747 airplanes and hasn't cracked yet.

One of the first people to walk on it was the original moonwalker, Buzz Aldrin.

On June 30, 1956, a Trans World Airlines and a United Airlines airplane both flying from Los Angeles to Chicago collided over uncongested airspace 21,000 feet over the Grand Canyon, killing all 128 people onboard the two flights. The collision became the leading factor in the founding of the FAA, America's Federal Aviation Administration.

Today, the park operates six no-fly zones and special flight rules to reduce collision risk and noise pollution.

Inversion

The Grand Canyon's most peculiar weather phenomena. Inversion is when fluffy thick clouds fill the entire Canyon below the rim.

A sight to see.

250

The annual average of people rescued from within the park. Approximately one every 1.3 days.

One of the main reasons for rescue, according to park rangers, is insufficient water and improper footwear. (Some tourists have been rescued wearing high heels and flip-flops.)

As of 2021, about 900 people have been recorded to have died while within the Grand Canyon National Park.

770 of those deaths were recorded between 1860 and 2015, including: 53 resulting from falls; 65 resulting from "environmental causes", such as heat stroke, cardiac arrest, dehydration and hypothermia; 7 caught in flash floods; 79 drowned in the Colorado River; 242 died in airplane and helicopter crashes; 25 died in freak accidents, including lightning strikes and rock falls; and 23 were victims of homicide.

In June 2014, American tightrope walker Nik Wallenda became the first daredevil to traverse the Grand Canyon on a wire. He left his safety harness at home.

He made it across the 1,400-foot-long, 8.5-ton cable suspended 1,500 feet above the Little Colorado River in just 23 minutes. He prayed out loud for the entire stunt.

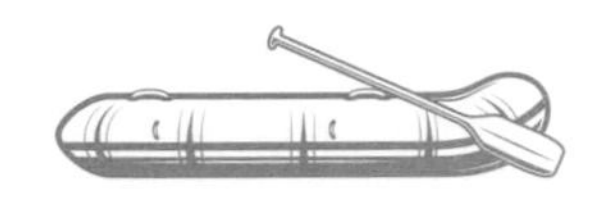

A canyon is described as a deep, narrow valley with steep sides*.

The word Canyon comes from the Spanish word *cañon*, translated as "tube" or "pipe".

**Never call the Grand Canyon a "gorge". It isn't – a gorge is steeper and narrower than a canyon. It is gorge-ous, of course.*

China's Yarlung Tsangpo is a gorge three times deeper than the Grand Canyon. Carved out of pure granitic bedrock, Yarlung Tsangpo hits the floor running at 5,300 metres (17,000 feet).

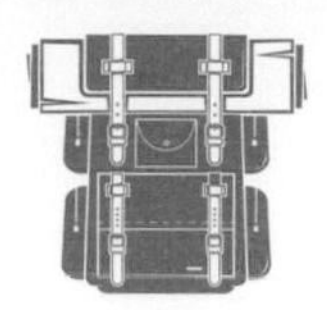

“You cannot see the Grand Canyon in one view, as if it were a changeless spectacle from which a curtain might be lifted, but to see it you have to toil from month to month through its labyrinths.”

John Wesley Powell

The Spanish explorer García López de Cárdenas and his Coronado Expedition team arrived at the Grand Canyon in 1540 but, as it turned out, they were not looking for one of the Seven Natural Wonders of the World. They were searching for the legendary Seven Cities of Cibola – the now-mythical lands thought to be filled with gold that Spanish explorers believed existed in the southwest of North America, comparable to the mythical city of El Dorado.

With the Grand Canyon, they found something even more priceless – tourism.

"

The landscape everywhere, away from the river, is of rock – cliffs of rock; plateaus of rock; terraces of rock; crags of rock – ten thousand strangely carved forms.

"

John Wesley Powell

According to the creation stories of the Navajo, the Grand Canyon was also created after a great flood. To save themselves, they transformed into fish that lived in the water until the water drained away, leaving the land in the shape of the great Canyon that we see today. The Navajo ancestors then transformed themselves back into human form.

According to the creation stories of the Havasupai, the Grand Canyon was created when the two ruling gods of the universe, Hokotama and Tochopa, started a civil war to fight for control over the world. Hokotama drowned the entire world in a huge flood, where every human died. However, Tochopa hid his daughter, Pukeheh, in a hollow tree and she repopulated the Earth with life again after the flood passed.

“Arizona is known for five ‘Cs’ – copper, citrus, cotton, cattle, and climate.”

Kathleen Derzipilski (she forgot the Canyon).

South Kaibab Trail

The number-one must-do hiking trail in the South Rim of Grand Canyon National Park. The South Kaibab Trail is the one path that goes down into the Canyon, crosses the Colorado River and comes back up the other side. It's a 21-mile hike, so an overnight stay in the Canyon is required to complete it.

(928) 638-2477

The number to call if you require a 24-hour emergency response within the park – if you can get reception (which you won't).

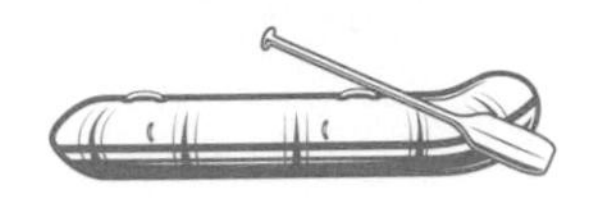

You can raft the entire 277 miles of the Colorado River portion of the Grand Canyon.

Using a powered raft will take about about 7 days, paddling will take a lot longer, between 15 days and 3 weeks!

In 2018, three 5-gallon paint buckets filled to the top with highly radioactive uranium ore were found – open – next to the taxidermy exhibit at the National Park's Museum Collection Building.

A 2019 report concluded that no radiation had been exposed as a health risk to employees or visitors over a period of 20 years. Uranium naturally occurs in the rocks of Grand Canyon National Park.

100 miles

The distance you can see atop the Grand Canyon on a clear day.

314.6 years

This is how long it would take to fill the Grand Canyon, with a volume of 4.17 trillion cubic meters, with water (at 420 cubic meters per second, the average flow rate of the Colorado River).

"Monumental is not a matter of size."

Albert Paley

"Standing on the edge of the Grand Canyon and contemplating your own greatness is pathological. At such moments we are made for a magnificent joy that comes from outside ourselves."

John Piper